PLANETA ANIMAL

LA IGUANA

POR VALERIE BODDEN

CREATIVE EDUCATION • CREATIVE PAPERBACKS

Publicado por Creative Education
y Creative Paperbacks
P.O. Box 227, Mankato, Minnesota 56002
Creative Education y Creative Paperbacks son marcas editoriales de The Creative Company
www.thecreativecompany.us

Diseño de The Design Lab
Producción de Rachel Klimpel
Dirección de arte de Rita Marshall
Traducción de TRAVOD, www.travod.com

Fotografías de Alamy (age fotostock, blickwinkel, Volodymyr Burdiak), Corbis (Christian Ziegler/Minden Pictures, ZSSD/Minden Pictures), Dreamstime (Adogslifephoto, David Morton, Edgar Barany, Isselee), iStock (alkir, Che_Tina_Plant), Minden Pictures (Tim Fitzharris), Shutterstock (apiguide, Kjersti Joergensen, Sergey Uryadnikov, Shackleford-Photography, Vikulin)

Library of Congress Cataloging-in-Publication Data
Names: Bodden, Valerie, author.
Title: La iguana / by Valerie Bodden.
Other titles: Iguana. Spanish
Description: Mankato, Minnesota: Creative Education/Creative Paperbacks, [2023] | Series: Planeta animal | Includes index. | Audience: Ages 6–9 | Audience: Grades 2–3 | Summary: "Elementary-aged readers will discover some iguanas can change colors. Full color images and clear explanations highlight the habitat, diet, and lifestyle of these fascinating creatures"—Provided by publisher.
Identifiers: LCCN 2022007740 (print) | ISBN 9781640266926 (library binding) | ISBN 9781682772485 (paperback) | ISBN 9781640008335 (ebook)
Subjects: LCSH: Iguanas—Juvenile literature.
Classification: LCC QL666.L25 B63518 2023 (print) | LCC QL666.L25 (ebook) | DDC 597.95/42—dc23/eng/20220228
LC record available at https://lccn.loc.gov/2022007740
LC ebook record available at https://lccn.loc.gov/2022007741

Tabla de contenidos

Algunas iguanas verdes son pardas (abajo), pero las iguanas crestadas de Fiyi (enfrente) son verdes.

Las iguanas son lagartos grandes. Como todos los lagartos, son reptiles. Los reptiles son animales **de sangre fría** con cuerpos escamosos. En el mundo, hay unos 40 tipos de iguanas.

de sangre fría animales cuyos cuerpos siempre están a la temperatura del aire que los rodea

La mayoría de las iguanas tiene un colgajo de piel bajo la barbilla llamado papada. Tienen colas largas y patas fuertes. Las iguanas tienen espinas en la espalda. La mayoría de las iguanas son verdes, grises, o del color café. Algunas pueden cambiar de color. Se oscurecen cuando están frías, y se aclaran cuando están calientes.

La papada de una iguana saludable es suave y puede moverse fácilmente.

La iguana más pequeña mide unas 10 pulgadas (25,4 cm) de largo. ¡Pero la iguana más larga mide siete pies (2,1 m), desde la nariz hasta la punta de la cola! Las iguanas más pesadas pueden pesar 30 libras (13,6 kg).

La iguana marina de las Galápagos puede llegar a pesar entre 20 y 30 libras (9,1–13,6 kg).

Las iguanas viven en América del Sur y del Norte. También viven en algunas islas. A las iguanas les gustan los lugares calurosos. La mayoría hace su casa en el bosque. Algunas viven en tierras calurosas y secas llamadas desiertos.

La iguana calienta su cuerpo bajo los rayos del sol, hasta que es demasiado.

La iguana se traga su alimento entero, sin masticar.

La mayoría de las iguanas comen frutas, flores y hojas. A veces, comen **insectos**. ¡Un tipo de iguana incluso se sumerge en el agua para comer algas marinas!

insectos animales pequeños con el cuerpo dividido en tres partes y que tienen seis patas

La hembra cava un nido donde pone todos sus huevos.

Algunas hembras pueden poner 70 huevos. Las iguanas bebés **eclosionan** todas al mismo tiempo. Viven en grupo durante un tiempo. Su madre no las cuida. A medida que crecen, las iguanas mudan de piel. La mayoría de las iguanas viven entre 15 y 20 años en la naturaleza.

eclosionar salir de un huevo

Sus patas con garras le ayudan a la iguana a trepar por ramas resbalosas.

Muchas iguanas pasan su tiempo en los árboles. ¡Pueden caer de una altura de 50 pies (15,2 m) sin lastimarse! A las iguanas les gustan los árboles con ramas que cuelgan sobre el agua. Pueden escapar de algún **depredador** dejándose caer en el agua. Usan sus colas fuertes para huir nadando.

depredador animal que mata y se come a otros animales

A veces, las iguanas se reúnen en grupos grandes. Se acuestan juntas a tomar sol. Pero los machos pueden pelearse entre sí por las hembras.

Comúnmente, las iguanas marinas forman los grupos más grandes.

Algunas personas tienen iguanas como mascotas. Pero puede ser difícil cuidarlas. Otras personas visitan a las iguanas en zoológicos o en la naturaleza. ¡Es divertido observar a estas lagartijas de cola larga que viven en los árboles!

Con una buena alimentación y cuidados, una iguana mascota puede vivir muchos años.

Un cuento de la iguana

En Norteamérica, se contaba una historia de por qué las iguanas viven en los árboles. Se decía que, un día, la iguana oyó un grito por ayuda. Pero lo ignoró y no ayudó a quien gritaba. Así que un dios la golpeó y la lanzó al río. La iguana salió del río y se escondió en un árbol. Allí ha vivido desde entonces.

dios un ser que se cree que tiene poderes especiales y control sobre el mundo

Índice

A Maker's Guide to

WHEELS

Written by JOHN WOOD

Illustrated by AMY LI

LERNER PUBLICATIONS ◆ MINNEAPOLIS

First American edition published in 2021 by Lerner Publishing Group, Inc.

Editor: Madeline Tyler
Design: Amy Li

Lerner Publications Company
An imprint of Lerner Publishing Group, Inc.
241 First Avenue North
Minneapolis, MN 55401 USA

For reading levels and more information, look up this title at www.lernerbooks.com.

Main body text set in VAG Rounded Std. Typeface provided by Adobe Systems.

Library of Congress Cataloging-in-Publication Data

Names: Wood, John, 1990– author. | Li, Amy, illustrator.
Title: A maker's guide to wheels / written by John Wood ; illustrated by Amy Li.
Other titles: Wheels
Description: First American edition. | Minneapolis : Lerner Publications, 2021. | Series: A maker's guide to simple machines | Includes index. | Audience: Ages 6–10 | Audience: Grades 2–3 | Summary: "What is a wheel, and why do we need it? Answer these questions and more in this title all about wheels. A project where readers make a toy car adds opportunity for hands-on learning!"— Provided by publisher.
Identifiers: LCCN 2020054238 (print) | LCCN 2020054239 (ebook) | ISBN 9781728416465 (lib. bdg.) | ISBN 9781728438351 (pbk.) | ISBN 9781728418919 (eb pdf)
Subjects: LCSH: Wheels—Juvenile literature.
Classification: LCC TJ181.5 .W66 2021 (print) | LCC TJ181.5 (ebook) | DDC 621.8—dc23

LC record available at https://lccn.loc.gov/2020054238
LC ebook record available at https://lccn.loc.gov/2020054239

Manufactured in the United States of America
1-48893-49207-12/17/2020

Table of CONTENTS

Words that look like <u>this</u> can be found in the glossary on page 24.

BE A MAKER

Oh no! Sorry about this mess—Maker can be a bit clumsy sometimes. Say sorry, Maker.

Believe it or not, Maker is very smart when it comes to machines.

A machine is an object that makes a job easier to do. Maker wants to teach you about one of the simplest types of machine: a wheel.

A wheelchair is a machine. So is a car.

WHEELS

Wheels are circular. They are often used to help vehicles move over the ground. Wheels come in many different sizes for many different jobs.

A wheel has a long, thin rod that goes through the middle. This rod is called an axle. The axle allows the wheel to spin.

Wheels aren't just on the ground, though. Here are some other examples of wheels in our world.

Wheels are inside lots of machines, too. They do all sorts of things.

PARTS OF A WHEEL

Maker says he wants to show you the parts of a wheel.

Some axles have bearings. Bearings let the wheel turn smoothly around the axle. Most bearings are made up of tiny metal balls that fit around the axle. These are called ball bearings, and they roll as the axle or wheel spins.

HOW A WHEEL WORKS

There is a part that makes the axle and wheel turn. This is different for every machine. On a bike, your feet and the pedals turn the wheels. In a car, the engine turns the wheels.

Axles and wheels can spin either clockwise or counterclockwise.

In many machines, the wheels are powered by electric motors.

It is much easier to turn the small axle instead of the whole wheel. It can make things move farther with a smaller amount of energy.

FRICTION

Friction is a force caused by two things rubbing against each other. Friction stops things from moving, and you need lots of energy to overcome it.

Every part of the bag touching the floor is causing friction. All that friction adds up.

Friction causes heat. It also causes things to be worn away over time.

Wheels help move things because they reduce friction. Only the bottoms of the wheels are touching the ground.

Less friction means things take less energy to move.

BETTER WHEELS

There are lots of ways to make wheels do more work for less energy. Usually this means reducing friction.

Putting grease or oil on the axle helps the wheel spin more smoothly. Grease and oil make things slippery.

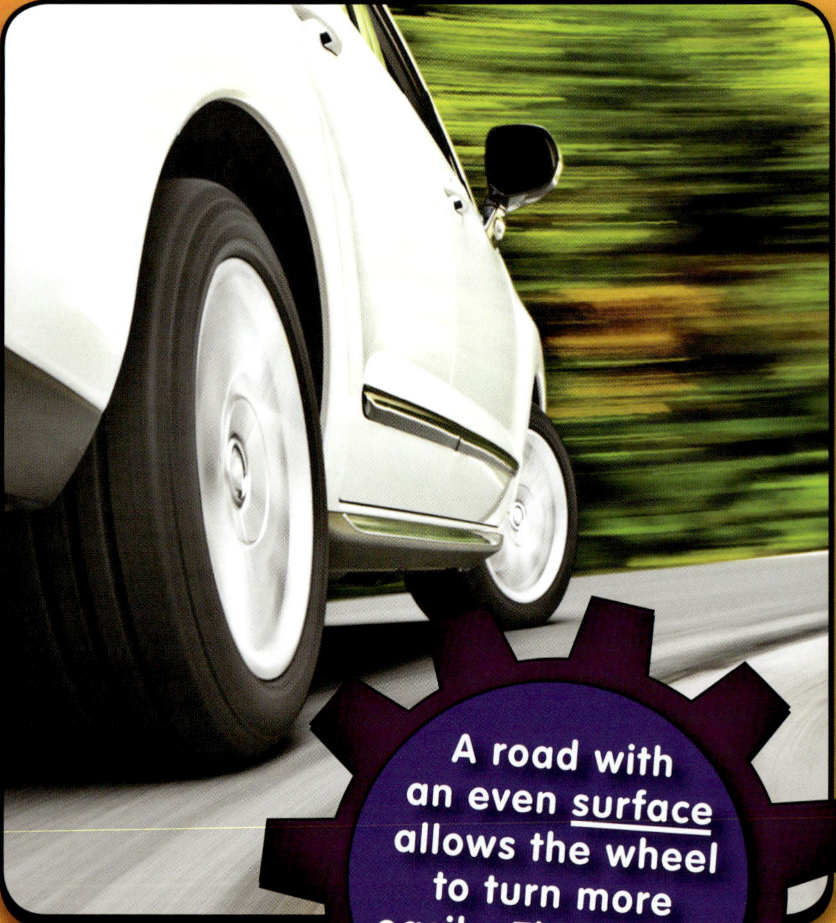

A road with an even surface allows the wheel to turn more easily. This makes it easier to drive on.

When a machine has big wheels, it means that the wheels work harder when the axle turns. However, wheels that are too big can cause all sorts of problems.

If the wheels on this racing car were too big, they would be heavy and hard to drive.

BUILD A TOY CAR

It is time to build! We will be using wheels to make a car that moves forward on its own. You will want an adult to help you with the cutting, sticking, and tying.

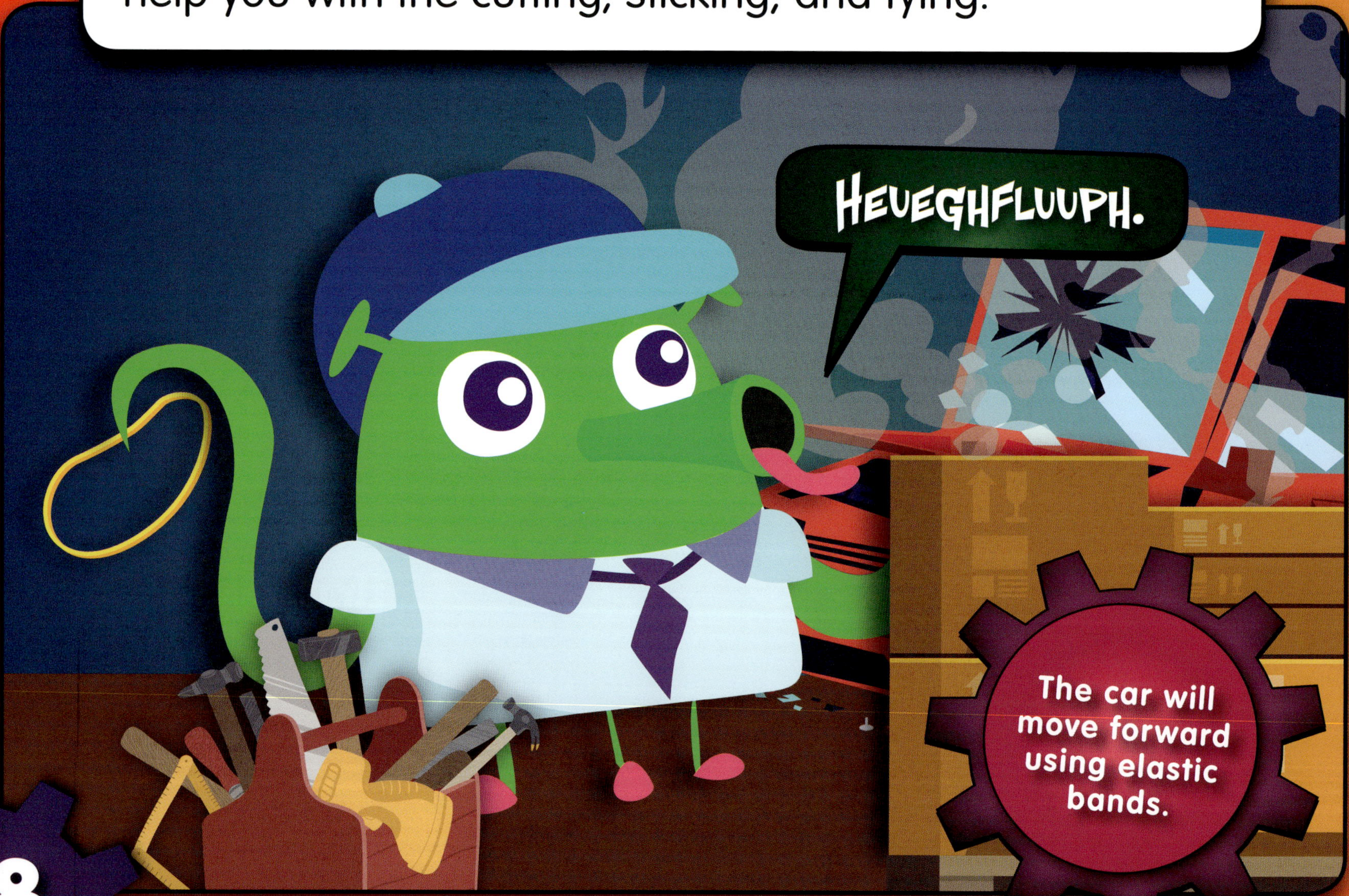

The car will move forward using elastic bands.

YOU WILL NEED:

- 2 straws
- Glue
- 2 straight, thin wooden sticks, 12 inches (30.48 cm) long
- 2 straight, thin wooden sticks, 2 inches (5.08 cm) long
- 3 toothpicks
- Scissors
- 4 wheels (you can use bottle caps with small holes in the middle)
- 3 elastic bands that are cut so they are flat
- An uncut elastic band

STEP 1

Glue one smaller stick between the longer sticks so they make a three-sided rectangle. This is the front. Glue the other smaller stick near the opposite end. This is the back.

STEP 2

Cut the straws so they are as long as the distance between the long sticks.

STEP 3

Glue a straw to each end of the long sticks. Make sure the straw is glued to both of the long sticks. Once dry, cut out the middle of the back straw.

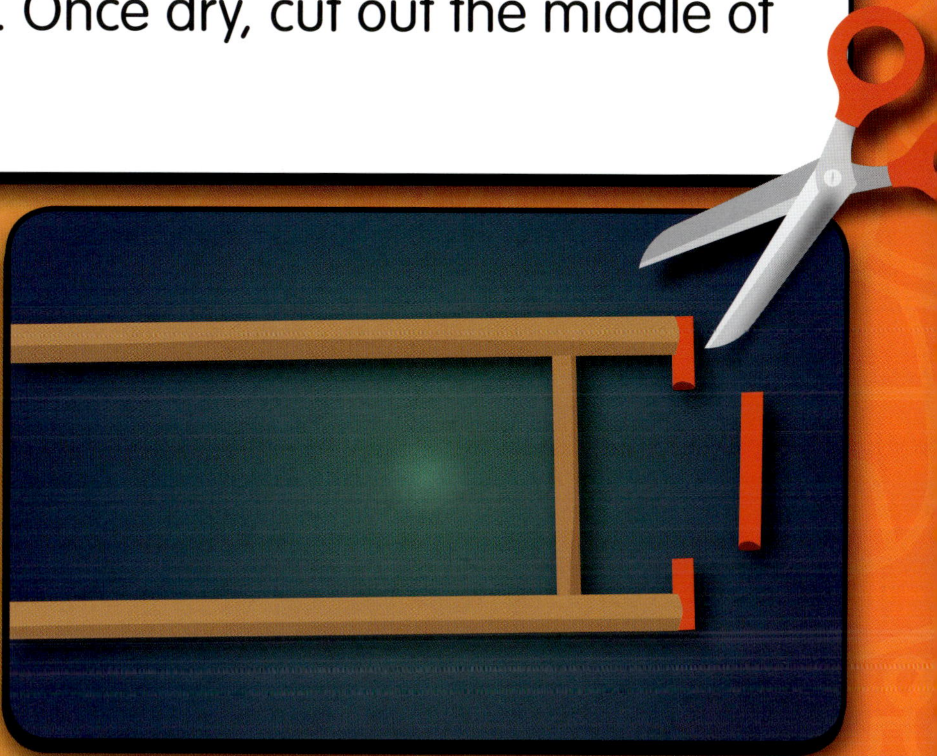

STEP 4

Slide a toothpick through each straw. These are your axles. Glue a wheel on each end of the toothpicks.

STEP 5

Cut out a very small part of the final toothpick and glue it so it sticks out of the back axle.

STEP 6

Tie your three cut elastic bands together. Tie one end to the front.

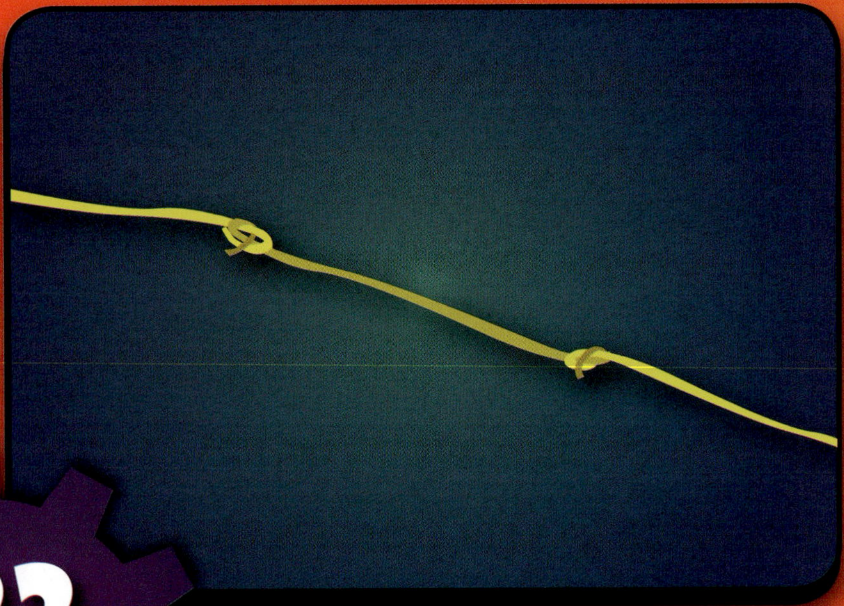

STEP 7

Tie the uncut elastic band to the end of the other bands. Then loop it around the tiny bit of toothpick on the back axle.

STEP 8

It's ready! Wind up the elastic bands by turning the back axle. Let go and watch it speed away!

GLOSSARY

electric motors	machines that move things using electricity
energy	a type of power that can be used
force	a push or pull on an object
overcome	to defeat, overpower, or deal with something
surface	the outside layer of something
vehicles	machines that are used to carry people or things

INDEX